Mouse Po

Compiled by John Foster

OXFORD

Oxford University Press, Walton Street, Oxford OX2 6DP

Oxford New York Toronto
Delhi Bombay Calcutta Madras Karachi
Petaling Jaya Singapore Hong Kong Tokyo
Nairobi Dar es Salaam Cape Town
Melbourne Auckland

and associated companies in
Berlin Ibadan

Oxford is a trade mark of Oxford University Press

© Oxford University Press 1991
Reprinted 1992
Printed in Hong Kong

A CIP catalogue record for this book is available from the British Library.

Acknowledgements
The Editor and Publisher wish to thank the following who have kindly given permission for the use of copyright material:

Mary Dawson for 'Ginger Cat' © 1990 Mary Dawson; Eric Finney for 'Mouse and Lion' © 1990 Eric Finney; John Foster for 'In the middle of the night' © 1990 John Foster; Jean Kenward for 'The House's Tale' © 1990 Jean Kenward; Irene Rawnsley for 'The Brave Mouse' and 'Two Mice' both © 1990 Irene Rawnsley; John Rice for 'A Mouse in the Kitchen' © 1984 John Rice.

Although every effort has been made to contact the owners of copyright material, a few have been impossible to trace, but if they contact the Publisher correct acknowledgement will be made in future editions.

In the middle of the night

In the middle of the night,
While we slept,
The mouse crept
Out of the nest
Beneath the floor boards.

In the middle of the night,
While everything was quiet,
The mouse scampered
Across the kitchen floor
Searching for breadcrumbs.

In the middle of the night
While Mum and Dad slept
I crept
Quietly down the stairs
To get myself a drink.

In the middle of the night
When I opened the door
Of the kitchen
I saw a flash of fur
As a small brown mouse
Shot past me.

And I jumped with fright
In the middle of the night.

John Foster

3

Jenny Williams 198

The House's Tale

This is the HOUSE
that was built in the road
where SAM lives.

These are the BRICKS
that made the house
that was built in the road
where SAM lives.

This is the STRAW
that packed the bricks
that made the house
that was built in the road
where SAM lives.

This is the LORRY
that carried the bricks
(all packed in straw)
that made the house
that was built in the road
where SAM lives.

4

This is the MOUSE
that lived in the straw
that packed the bricks
that filled the lorry
that drove to the house
where SAM lives.

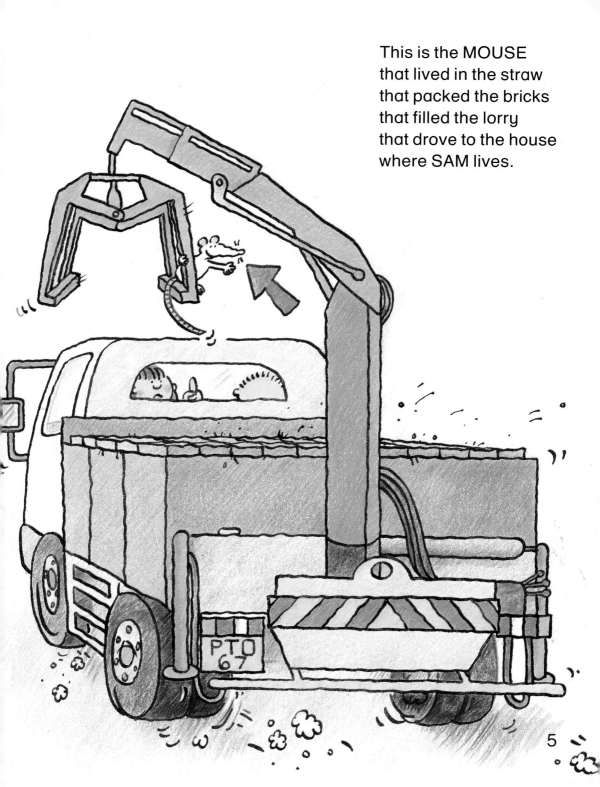

He jumped and jittered
from paw to paw,
he hurried and scurried. . ..
he peeped, and saw
a roof, a window,
an open door
and tables and chairs,
and on the floor
some biscuit crumbs
and an apple core
in the house in the road
where SAM lives.

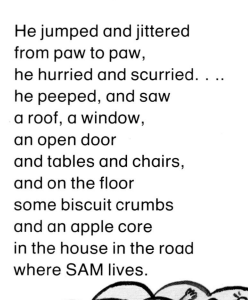

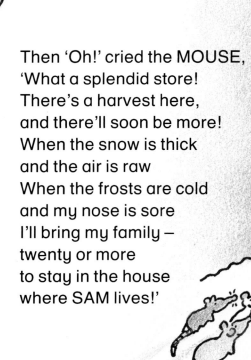

Then 'Oh!' cried the MOUSE,
'What a splendid store!
There's a harvest here,
and there'll soon be more!
When the snow is thick
and the air is raw
When the frosts are cold
and my nose is sore
I'll bring my family –
twenty or more
to stay in the house
where SAM lives!'

6

So the MICE arrived
one winter's night
when the moon was full
and the stars were bright
and the Christmas angels
took their flight.
They all crept in
from left and right
to stay in the house
where SAM lives.
Did ever you see
such a splendid sight?
Just LOOK at the HOUSE –
the family house –
that was built in the road
where SAM lives!

Jean Kenward

7

Ginger Cat

Sandy and whiskered, the ginger cat
Sniffs round the corners for mouse or for rat;
Creeping right under the cupboard he sees
A little mouse having a nibble of cheese.

On velvety paws with hardly a sound
The ginger cat watches, and padding around
He finds a good hiding-place under a chair,
And sits like a statue not moving a hair.

Then baring his claws from their velvety sheath
He pounces, miaowing through threatening teeth.
But puss is too late, for the sensible mouse
Was eating the cheese at the door of his house.

Mary Dawson

A Mouse in the Kitchen

There's a mouse in the kitchen
    Playing skittles with the peas,
He's drinking mugs of coffee
    And eating last week's cheese.

There's a mouse in the kitchen
    We could catch him in a hat,
Otherwise he'll toast the teacakes
    And that's bound to annoy the cat.

There's a mouse in the kitchen
    Ignoring all our wishes,
He's eaten tomorrow's dinner
    But at least he's washed the dishes.

John Rice

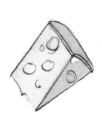

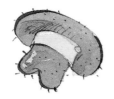

Mouse and Lion

Mouse caught by lion
Pleads for his life,
Begs to go home
To his children and wife:
'You never know, Leo,
If you set me free
One day you might get
In a fix and need me.'
Leo, amused by
The fieldmouse's cheek,
Lets him go. He departs
With a skip and a squeak
And, would you believe it,
The very next week
Leo the lion
Just happens to get
Hopelessly tangled in
Game hunters' net.
Mouse happens along,
Says, 'I'm small, with no roar,
But there's one thing I can do,
And that is to gnaw.'
In less than an hour
The net's nibbled through. . .

Do someone a good turn,
He might do one for you.

Eric Finney

Two Mice

Two mice lived
In a garden wall;
She made a warm nest,
He found a hole.

One wore a grey coat,
The other a brown;
She liked country,
He liked town.

He ate bacon,
She ate barley;
He slept late
But she woke early.

14

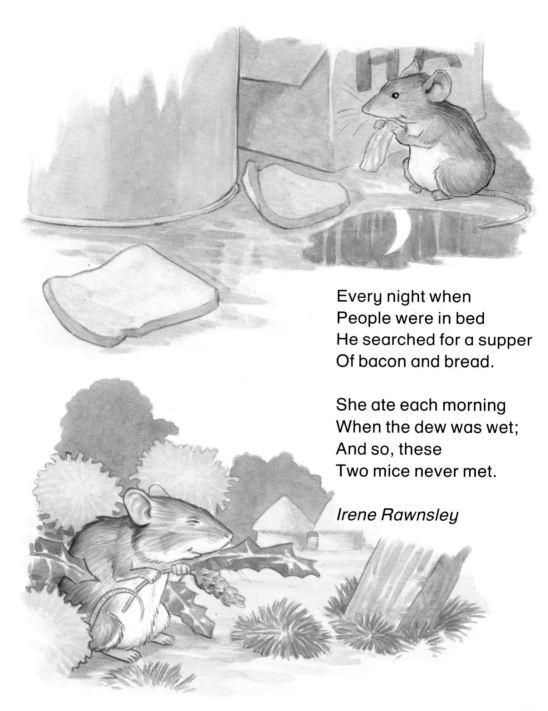

Every night when
People were in bed
He searched for a supper
Of bacon and bread.

She ate each morning
When the dew was wet;
And so, these
Two mice never met.

Irene Rawnsley

The Brave Mouse

Annabel,
Annabel,
Come and see here;
A mouse is asleep
In tabby cat's ear!

He climbed up her tail
As she lay in a heap;
Ran over her body
Then fell fast asleep.

I wonder
I wonder
For brown mouse's sake
If he or if tabby
Will be first awake?

Irene Rawnsley